POINTS OF VIEW

Is TECHNOLOGY Becoming Too Smart?

By Theo McDunn

KidHaven PUBLISHING

Published in 2022 by
KidHaven Publishing, an Imprint of Greenhaven Publishing, LLC
353 3rd Avenue
Suite 255
New York, NY 10010

Designer: Deanna Paternostro
Editor: Caitie McAneney

Photo credits: Cover Tanoy1412/Shutterstock.com; p. 5 Flashpop/DigitalVision/Getty Images; p. 7 Roslan Rahman/Contributor/AFP/Getty Images; p. 9 (top) Jasmin Merdan/Moment/Getty Images; p. 9 (bottom) JGI/Jamie Grill/Getty Images; p. 11 Laura Olivas/Moment/Getty Images; p. 13 Marko Geber/DigitalVision/Getty Images; p. 15 Antonio_Diaz/iStock/Getty Images Plus/Getty Images; p. 17 Teera Konakan/Moment/Getty Images; p. 19 valentinrussanov/E+/Getty Images; p. 20 SOPA Images/Contributor/LightRocket/Getty Images; p. 21 (notepad) ESB Professional/Shutterstock.com; p. 21 (markers) Kucher Serhii/Shutterstock.com; p. 21 (photo frame) FARBAI/iStock/Thinkstock; p. 21 (inset, left) Klaus Vedfelt/DigitalVision/Getty Images; p. 21 (inset, middle-left) Ariel Skelley/DigitalVision/Getty Images; p. 21 (inset, middle-right) Hispanolistic/iStock/Getty Images Plus/Getty Images; p. 21 (inset, right) wellphoto/iStock/Getty Images Plus/Getty Images.

Library of Congress Cataloging-in-Publication Data

Names: McDunn, Theo, author.
Title: Is technology becoming too smart? / Theo McDunn.
Description: New York : KidHaven Publishing, 2022. | Series: Points of view | Includes bibliographical references and index.
Identifiers: LCCN 2020030745 | ISBN 9781534536470 (library binding) | ISBN 9781534536456 (paperback) | ISBN 9781534536463 (set) | ISBN 9781534536487 (ebook)
Subjects: LCSH: Technology–Juvenile literature. | Artificial intelligence–Juvenile literature.
Classification: LCC T48 .M564 2022 | DDC 303.48/3–dc23
LC record available at https://lccn.loc.gov/2020030745

Printed in the United States of America

Some of the images in this book illustrate individuals who are models. The depictions do not imply actual situations or events.

CPSIA compliance information: Batch #CS22KH: For further information contact Greenhaven Publishing LLC, New York, New York at 1-844-317-7404.

CONTENTS

Smarter All THE TIME

Technology is getting smarter all the time! Devices can do more, and they can even act on their own. Today, most homes in the United States have smartphones and other smart devices, or machines connected to the internet. Smart robots can clean public places, help doctors, and work in factories. Smart technology has changed the way we meet, shop, and learn. But is this technology going too far?

Some people think that technologies such as **artificial intelligence** (AI) are getting too smart. Other people think smart technologies can help us learn, create, and stay healthy. It's important to learn about both sides before you decide where you stand.

Know the Facts!

The Pew Research Center found that 90 percent of adults in the United States used the internet in 2019.

Many kids and teens use smartphones, which are connected to the internet, to chat with friends. It's important to learn smartphone and online safety!

How Do People Use TECHNOLOGY?

From making fire to using wheels, technology has changed life for humans for hundreds of thousands of years. Humans **developed** computers and the internet pretty recently in history. Since then, we've created new and related technologies very quickly.

Today, businesses not only use technology to make goods, they also use it to sell them online. People use smart devices in their homes to **regulate** heat, lock the doors, and turn lights on and off. Video meetings sometimes replace in-person meetings. Both sides of the debate, or argument, about smart technology agree that it has become a big part of society.

Know the Facts!

In 2019, around 53 percent of children in the United States owned a smartphone by the age of 11.

Robots like Spot the robot dog can do jobs such as helping police keep people apart in public during a **pandemic**.

Helpless HUMANS

Some people argue that as technology gets smarter, humans are actually becoming more helpless. They say humans depend on technology instead of learning and practicing basic skills.

For much of human history, people had to take care of many basic needs. Now, "smart homes" can regulate temperatures, turn on lights, and preheat ovens. Smartphones and **virtual** assistants, or helpers, remind you of events. Smartphone calculators can help with math homework. Online maps help you find your way. You might have all the knowledge in the world at your fingertips, but is it making you any smarter? Some people say no!

Know the Facts!

People can **hack** into smart homes and **security** systems to cause trouble or record personal images.

Many people argue that technology draws a line between people who have it and people who don't. Some people have smart homes, while others can't even **access** the internet for homework.

Learn Anything, ANYWHERE!

Other people think that technology is a useful learning tool. With internet access, kids and adults can look up facts, read books, and watch helpful educational videos. In this way, the internet and smart devices can make learning more equal. Anyone with the internet can study a subject, whether it's for school or just for fun.

Technology can help students with different learning styles. Students can use smart technology to learn in different ways within the same classroom. Teachers can use recorded lessons and online videos, **interactive** whiteboards, and hands-on smart devices to teach in different ways too.

Know the Facts!

The Pew Research Center found that around 58 percent of eighth-grade students used the internet every day for homework in 2018.

The COVID-19 pandemic hit the United States in 2020. Most schools had to switch to distance learning—learning at home—for a while. Devices such as laptops and tablets were useful tools for distance learning.

Privacy, PLEASE!

If smart technology can track how you move, what you buy, and what your habits are, do you have any privacy? Some people argue that smart technology knows too much about us.

Virtual assistants or smart speakers such as Google Home and Amazon Echo "listen in" to people's requests. They keep track of their owner's location, their purchases, and their tastes. Then, companies use this information to sell products. Sometimes, companies share this information, either on purpose or if they're hacked. Some people believe the privacy concerns aren't worth the benefits of smart technology.

Know the Facts!

By 2020, nearly 1 in 4 American adults owned a smart speaker.

In some families, smart speakers have become part of the household. People can ask the speaker to play a bedtime story or music for a dance party.

Health HELPERS

Some people argue that smart technology can save lives. Many people wear smartwatches, which can track their heart rate and movement throughout the day. That may push them to be more active and watch their health.

New AI technology can measure blood flow. Someday, it could be used to warn of a heart attack or stroke before it happens. Tiny robots called nanorobots are being developed to find and get rid of things that cause illness. Robots help in hospitals, even with operations. Drones, or flying robots, can drop off medicines and supplies to hard-to-reach places.

Know the Facts!

In 2018, technology **experts** were asked how they think AI will change human lives in the future. Many said smart technology could save time, money, and lives.

Many smartwatches work as fitness trackers. They keep track of a person's movement, heart rate, sleep, and other things.

Stealing JOBS

Smart technology can help in every field, including **manufacturing** and education. Some argue that the smarter technology gets, the more jobs it will take away from human workers. Virtual assistants can take the place of human assistants. AI can use information gathered through smart devices to **advertise** for a company.

When robots first started working in factories, they needed people to control them. Now, AI allows some robots to work on their own. In 2019, Oxford Economics—a group that studies business—said that up to 20 million manufacturing jobs could be lost to robots by 2030.

Know the Facts!

Oxford Economics said that **repetitive** jobs are most likely to be taken by robots. Jobs that require creativity and social skills are safer from robots.

Some people are worried that robots will take jobs away from people who don't have many other skills or opportunities.

Work Smarter, NOT HARDER

Some people think smart technology will help humans work smarter, not harder. They argue that AI can help people do things more efficiently, or better in less time. Virtual assistants help people plan their calendars and keep track of to-do lists. With smartphones, people can check emails, have video meetings, and do work from almost anywhere!

Since AI often takes over repetitive, boring tasks, some people think human workers will be able to do more exciting work. Instead of working in manufacturing, for example, people could train for other jobs that are more social or creative.

Know the Facts!

In a 2018 study of technology experts, 63 percent of them believed most people would be better off using technology with artificial intelligence in 2030.

Some people think smart technology only improves our ability to work and be creative.

A Smart DEBATE

Smart technology isn't going anywhere. It's on the rise, becoming a part of our everyday lives. Some people think it should be regulated so it doesn't become too powerful or harmful. Others say smart devices are worth the possible problems for the sake of efficiency and ease.

Now that you've heard both sides of the debate, what do you think about smart technology? When do you think it's useful? Could it ever be harmful? New technology will affect your life, so it's important to stay smart about it!

Know the Facts!

Elon Musk, who runs Tesla and SpaceX, has said he believes that AI should be regulated both by country and worldwide.

Is technology becoming too smart?

YES

- People rely on smart technology instead of gaining basic life skills.
- Smart technology breaks people into two groups—those who can afford it and those who cannot.
- Using some smart devices means a loss of privacy.
- Smart technology can take jobs away from human workers.

NO

- Smart technology is a useful learning tool.
- Smart technology makes us more equal, because access to the internet means anyone can learn anything.
- Smart technology can save lives and make people healthier.
- Smart technology can make us work more efficiently.

Use your experiences with smart technology, along with facts and opinions from both sides, to form your educated argument!

GLOSSARY

access: To use or have something. Also, the ability to use or have something.

advertise: To call attention to a product to make people want to buy it.

artificial intelligence: An area of computer science that deals with giving machines the ability to think and act like humans.

develop: To cause something to become more advanced.

expert: Someone who has a special skill or knowledge.

hack: To use a computer to gain unapproved access to data in a system.

interactive: Involving the actions or input of a user.

manufacturing: The process of making products by hand or by machinery.

pandemic: An outbreak of disease that occurs over a wide area and affects a large number of the population.

regulate: To fix or adjust something. Also, to bring order and rule to something.

repetitive: Happening again and again.

security: Having to do with the quality of being safe from danger.

technology: A method that uses science to solve problems and the tools used to solve those problems.

virtual: Having to do with being on a computer or online.

For More
INFORMATION

WEBSITES

Online Safety
kidshealth.org/en/kids/online-id.html
Learn how to stay safe and protect your privacy while you're using smart devices.

Robots
www.brainpop.com/technology/computerscience/robots/
Learn about robots through a movie, quiz, and other tools with BrainPop!

BOOKS

Dickmann, Nancy. *Robots and Artificial Intelligence*. New York, NY: Gareth Stevens Publishing, 2020.

Leigh, Anna. *Cutting-Edge Artificial Intelligence*. Minneapolis, MN: Lerner Publications, 2019.

Schuette, Sarah L. *Online Safety*. Mankato, MN: Capstone Press, 2020.

INDEX